Life and Technology

MW00795087

After Simondon Series

Edited by Erich Hörl and Yuk Hui

Jean-Hugues Barthélémy is Director of the Centre international des études simondoniennes (Maison des Sciences
de l'Homme Paris-Nord / Fondation "pour la science"),
editor and director of the *Cahiers Simondon*, and associated
researcher at the laboratory EA 4414 HAR (University Paris
Ouest—Nanterre La Défense). He is the author of several
monographs on Simondon, including his recent *Simondon*
(Paris: Les Belles Lettres, 2014; translation forthcoming:
Bloomsbury, 2016), and of many articles on contemporary
French and German philosophy.

Life and Technology: An Inquiry Into and Beyond Simondon

Jean-Hugues Barthélémy

Translated by
Barnaby Norman

μ meson press

**Bibliographical Information of the
German National Library**
The German National Library lists this publication in the
Deutsche Nationalbibliografie (German National Biblio-
graphy); detailed bibliographic information is available
online at http://dnb.d-nb.de.

Published in 2015 by meson press, Hybrid Publishing Lab,
Centre for Digital Cultures, Leuphana University of Lüneburg
www.meson-press.com

Design concept: Torsten Köchlin, Silke Krieg
Copy-editing: Damian Veal

The print edition of this book is printed by Lightning Source,
Milton Keynes, United Kingdom.

ISBN (Print):   978-3-95796-070-2
ISBN (PDF):   978-3-95796-071-9
ISBN (EPUB):   978-3-95796-072-6
DOI: 10.14619/015

The digital editions of this publication can be downloaded
freely at: www.meson-press.com.

Funded by the EU major project Innovation Incubator
Lüneburg

Contents

After Simondon Series Preface

Thanks largely to the works of philosophers who are inspired by him, most notably Gilles Deleuze and Bernard Stiegler, the name Gilbert Simondon is becoming more and more familiar to readers outside France. Up to the time of writing this preface, however, few of his works have been translated into English. It is almost an irony that we call this book series *After Simondon*, dedicated as it is to a thinker who is not yet fully available to his readers. However, *After Simondon* does not mean to overtake Simondon by declaring his thought obsolete, but rather to address him as our contemporary. Indeed, there are challenging contemporary issues that Simondon did not and could not address in his time, yet which his thought retains the power to interrogate, problematize, critique and illuminate.

This book series traces the implications as well as the critiques of Simondon's thought. It aims to go one step further than simply resituating Simondon as a neglected great twentieth-century philosopher of technology. Simondon was not merely a philosopher of technology but rather one whose ambition was nothing less than to rewrite the history of philosophy according to the concept of individuation and to invent a philosophical thinking that could effectively integrate technology into culture. *After Simondon* thus poses the question: What could critical thinking and theory concerning technology and individuation be *after* Simondon—that is, both *following* Simondon but also *going beyond* him and transgressing his thought?

We contend that Simondon's concepts and observations could serve as a rich source for the development of new concepts, theories and practices for coping with our contemporary condition. This includes a wide range of topics from digital objects and techno- and media-ecologies to what might be called a 'technological humanism'; from individuation, inventions and imaginations to perceptions; from animals to technical systems; and from issues of the automatic and alienation in the

10   twenty-first century to the process of cyberneticization. We hope
      that this series can act as a continuation of Simondon's projects,
      and we welcome proposals from scholars who are working on
      such subjects in relation to Simondon's thought.

                                        Erich Hörl and Yuk Hui
                                        Summer, 2015

Author's Preface to the English Translation

The texts brought together here were first published in French in two contributed volumes, edited respectively by Jean-Claude Ameisen and Laurent Cherlonneix, and by the late Jean-Marie Vaysse.[1] Erich Hörl and Yuk Hui had the idea of selecting these two texts to inaugurate the series *After Simondon*, and I thank them warmly for this. My aim is to provide the reader with a rigorous presentation of some of Simondon's key ideas, along with some developments that we can today bring to them.

Indeed, these two texts share a double ambition. On the one hand, to analyse the general—and in my view the most profound—logic of what I refer to in my work as Simondon's "genetic encyclopaedism." And, on the other, to lead beyond Simondon, in the direction of that comprehensive but open (because anti-dogmatic) system on which I am working at the moment, and for which the concluding part of the second text establishes some strictly architectonic principles. In this respect, I would like to congratulate Barnaby Norman for his work of translation. Philosophical language is, we say in French, "a language in a language [*une langue dans une langue*]," and Barnaby Norman was able to convey this philosophical language into the English version.

1   Jean-Claude Ameisen and Laurent Cherlonneix, eds., *Nouvelles représentations de la vie en biologie et philosophie du vivant* [New Representations of Life in Biology and the Philosophy of the Living Being] (Brussels: De Boeck, 2013); Jean-Marie Vaysse, ed., *Technique, monde, individuation: Heidegger, Simondon, Deleuze* [Technics, Life, Individuation: Heidegger, Simondon, Deleuze] (Hildesheim: Georg Olms Verlag, 2006).

Aspects of a Philosophy of the Living

As I often do, I am going to try to explore here the theoretical potentialities, and hence the possible currency, of Gilbert Simondon's (1924–1989) work. To speak of potentialities is of course to recognize that Simondon did not *conceptualize* the simple *intuitions* that were his. Particularly since his texts very often seem to draw on philosophical theories (on the living being, the theories of Canguilhem and Bergson, and sometimes even Nietzsche) and scientific theories (on the living being, Simondon cites Rabaud) that are difficult to square with what, thanks to scientific progress, we know today of the various realities about which these theories made their assertions. But beside the fact that the *goals* Simondon had in mind when he ventured into this territory may themselves seem very topical—such as his intention to challenge the "anthropological break" too often accepted by philosophers in the name of what is "proper to the human being"—it must also be noted that the *tensions* found in Simondon's text come from the presence, alongside a superseded theoretical heritage, of genuine *idiosyncratic intuitions* which may themselves be conceptualized today. This is particularly true, as we will see, for his precursory and incomplete questioning of the concept of "information," which he argued *from very early on* would become *central*, and whose theoretical inadequacy he at the same time denounced—pre-empting on this second point the more recent reflections of Henri Atlan, who now makes reference to him.[1]

If, therefore, his work is today enjoying a resurgence of interest, even internationally, it is because his questioning and his intuitions have a possible currency, whose force and extension I have been attempting to expose for ten years.[2] To the subject of the *living being*, along with the *non*-living and *psycho-social* life, Simondon brings a mode of questioning that does not exactly belong to his epoch, but whose initial strangeness makes more sense today.

The Positioning of the Thinking of the Living Being at the Centre of Genetic Encyclopedism

For Simondon, the living being is *simultaneously*:
– the object that is the most difficult to think; and
– the theme that contains the hidden unity of his work, even beyond that first surface unity presented by the theme of individuation, which is actually transversal for him.

1   Henri Atlan, *Le vivant post-génomique, ou Qu'est-ce que l'auto-organisation?* [Post-Genomic Living, or What is Auto-Organization?] (Paris: Odile Jacob, 2011). There will be an opportunity to talk about Atlan's Simondonian evolution below.

2   On the encyclopedic aspect of Simondon's approach, I refer to my overview of Simondon's work *Simondon ou L'Encyclopédisme génétique* (Paris: PUF, 2008). For a more technical examination of questions specifically connected with the living being, see Chapter IV of my book *Penser l'individuation: Simondon et la philosophie de la nature* (Paris: L'Hartmattan, 2005), as well as the two articles cited below by Anne Fagot-Largeault and Victor Petit. Simondon's thinking of the living being has received very little commentary, but these two articles are some of the best available in the field of exegetic work on Simondon's thought in general.

These are the two general points that I would like to quickly clarify by way of introduction to the more specific questions concerning biological theory that will be at issue in what follows.

First, then, the living being is the object that is the most difficult to think for Simondon. This is to be understood in two senses: a sense indicating an objective situation that Simondon lived through but did not think, and a sense that belongs to Simondon's own thought. So, on the one hand, Simondon lived through the objective situation of the biology of his epoch: in 1957, the year in which his crucial theoretical effort drew to a close,[3] the impermeability of the *germ cell* had of course been known about for more than half a century, but the double helix structure of DNA had only been known to biologists for four years—Simondon for his part only mentions Gesell's citation of "Wrinch's theory according to which the chromosome is a *structure* composed of two elements"[4]—and Crick was still several months off setting out what he would refer to as "the central dogma of molecular biology," which is to say, that the sense of genetic expression is univocal and that each gene has a corresponding transcript and protein. In France more than elsewhere, the debate between the *neo*-Darwinism deriving from August Weismann and *neo*-Lamarckism—which is to say between a more subtle Lamarckism and a Darwinism that was less Lamarckian than Darwin![5] —was

3   In *L'individuation à la lumière des notions de forme et d'information*, which was his main thesis for the *doctorat d'Etat*, supervised by Jean Hyppolite. Two works developed out of it, *L'individu et sa genèse physico-biologique* (Grenoble: Millon, 1995) (with a first incomplete edition published by PUF in 1964) and *L'individuation psychique et collective* (Paris: Aubier, 1989). The classic work *Du mode d'existence des objets techniques* (Paris, Aubier, 1958) was his secondary thesis.

4   Gilbert Simondon, *L'individuation à la lumière des notions de forme et d'information*, 207.

5   It gives me pleasure to recall here what Jean Gayon said about Darwin at the end of his famous study: "As for his theory of heredity, it was in general extremely obscure, and when it was clear, it was a manifesto for a an extreme form of the heredity of acquired characteristics" (*Darwin et l'après-Darwin* [Kimé, Paris: 1992], 411).

still going strong. Simondon made reference to Darwin and Lamarck, but in order to discuss their respective concepts of "adaptation" in remarks dedicated to the *philosophical* presup-positions of the biological debate, remarks which therefore remained relatively exterior to contemporary discussions on the innate and the acquired, with these two notions barely making an appearance in his text. For all that, it is possible to argue, with Anne Fagot-Largeault, that Simondon's position represents the invention of a "*technical* neo-Lamarckism,"[6] to the extent that Simondon wanted to think the living being *such that it engenders technics* and such that it defines (*via* the "process of hominization" that is the human being for Leroi-Gourhan) an inherited technical *world* which *appeals to* our various potentials—which, moreover, are *inextricably* individual and collective at the *psycho-social* level of the living beings that we are.

On the other hand, Simondon's thought itself makes the living being the object that is the most difficult to think: being a second "order of individuation" after the physical order, the living being is not, for all that, a *substantial domain* which would vindicate vital-ism. Simondon, like Georges Canguilheim, draws here on Claude Bernard's theoretical position from the *Introduction à l'étude de la médecine expérimentale* [Introduction to the Study of Exper-imental Medicine], a position—not however *theorized as such* by Bernard, who was relatively unconcerned in this respect—which

6   Anne Fagot-Largeault, "L'Individuation en biologie," in Bibliothèque du
     Collège international de philosophie, *Gilbert Simondon: Une pensée de
     l'individuation et de la technique* (Paris: Albin Michel, 1994) (my emphasis).
     Here applied to Simondon, the expression is taken by Fagot-Largeault from
     M. Tibon-Cornillot, whose article she cites, "Penser en amont de la bio-
     éthique: transformations dirigées du génome et crise du néodarwinisme," in
     Vers un anti-destin? Patrimoine génétique et droits de l'humanité, ed. François
     Gros and Gérard Huber (Paris: Editions Odile Jacob, 1992), 127–46. The idea of
     a specifically *technical* neo-Lamarckism has been developed—in extremely
     complex ways which I have discussed elsewhere—by Bernard Stiegler in the
     three volumes of *La Technique et le Temps* published to date (Paris: Galilée,
     1994, 1996 and 2001).

overcomes the opposition between mechanism and vitalism.[7] The
inherent difficulty of this enterprise—avoiding mechanism without then falling back into vitalism—is heightened by the fact that,
for Simondon, the living being must be thought of as *that which
makes possible a third order of individuation, simultaneously internal to the living being itself while extending and exceeding it*: the
psycho-social or "transindividual" order of individuation. Vitalism
is in fact even harder to avoid when your intention is to make the
living being something that is capable of becoming *psycho-social*.
But this intention is the necessary counterpart to the intention,
central to Simondon's work, of thinking man himself as a living
being. We will see that it is not possible to understand Simondon's discussion of the living being without seeing it in the light of
this exigency: to make culture emerge *from nature itself*. Further,
it will become evident that Simondon balances the "vitalist risk"
inherent to the way he would like to understand the genesis of
the psycho-social with the symmetrical ambition of deriving the
living from that which is not living. *Such a compensation will, however, produce the extreme theoretical difficulty of a "great division,"
which will nevertheless necessarily define Simondon's undertaking,
itself necessary, as the non-scientific—because philosophical—unification of the sciences, which in fact lack unity.*

We come now to the second of the general points—the theme of
the living being contains the hidden unity of Simondon's work,
even beyond that first surface unity presented by the transversal
theme of individuation. Indeed, the two essential works, *L'individuation à la lumière des notions de forme et d'information* [Individua-
tion in the Light of the Notions of Form and Information] and *Du
mode d'existence des objets techniques* [On the Mode of Existence

7   Discussion of this theoretical position taken by Claude Bernard, as well as
    the un-theorized tension it produces between the first two of the three parts
    of his major work, can be found in François Dagognet's very lucid preface to
    Introduction à l'étude de la médecine expérimentale (Paris: Flammaron, 1984).

of Technical Objects][8] do not only complement one another at the heart of a "Genetic Encyclopedism" (this being what this philosophy is called) which aims to think the individuation of physical, vital, psycho-social and technical beings. They are also articulated with each other within a constant dialogue with cybernetics, whose tendency to reduce the living being to technical schemas is criticized by Simondon. For Simondon, it is instead a matter of thinking the "concretization" of technical objects as an "individualization" for which the living being provides the model, which is only ever approached by the technical object in its relation with its "associated milieu." If, therefore, there are for Simondon "phylogenetic lineages" of technical objects, the analogy between the living being and the machine is not for all that an assimilation of the first to the second, and the machine is only made possible as something that functions because it is itself the work [œuvre] of a living being. So, Simondon's thought finds its general structure in an analogy which is not an identity between the technical and the living.

Now, the theme of "individualization" which Simondon transfers from a thinking of the living to a thinking of the technical object will at the same time provide the major idea of his thinking of the living being, insofar as individualization, as distinguished from what Simondon refers to as "individuation," is not only a genesis, but a *continual* genesis. This is in fact a possible first definition of life: the living, as distinguished from the physical, maintains its own becoming in terms of an individuation understood as a genesis. I will need to clarify this before coming *by this route* to the question of "adaptation," and then, by way of the question of *information*, to its possible relation to the question of apoptosis.

8   We know that it is through *Du mode d'existence des objets techniques* that Simondon became well known, but it is also through this work that he is mistakenly *reduced* to the status of a thinker of technics.

Individuation and Individualization: Life as Continual Genesis

I said that Simondon's thinking of the living being only makes sense in the light of his central challenge to the philosophers' "anthropological break." It is because the human being must be understood as a living being that life must be understood as potentially the bearer of a psycho-social becoming. This is the meaning of this strange formula, used by Simondon to denounce the philosophers' procedure: "you certainly cannot make the human being emerge from the vital if you extract the Human Being from the vital."[9] Mechanism, when applied to the living being, serves the interests of an initial anthropocentrism, which it is a question of challenging by returning to the living its ability to engender the human being and his spirituality. This double theoretical move is certainly not completely obligatory—you find theoreticians today, often biologists, who think the human being starting from the living being without, for all that, retaining the requirement of then making life capable of spirituality: for them, the "psycho-social" is nothing but an epiphenomenon, and humanity's most significant achievements only expressions of the struggle for survival! It is not by chance that Simondon was so interested in *ethology*. Ethologists, as specialists in animal behavior and its psychic dimension, are in fact best placed to challenge the cultural application of Darwinism, as the great ethologist Frans de Waal has done in his overview of the subject, *L'Âge de l'empathie* [The Age of Empathy].[10]

And yet I also emphasized that Simondon seeks to avoid a fall back into vitalism. How does he do it? By thinking human individualization as a "personalization" placed above two initial forms of individualization of the living being, *themselves rooted in a*

9   Simondon, *L'individuation psychique et collective*, 181, and *L'individuation à la lumière des notions de forme et d'information*, 297.

10   Frans de Waal, *The Age of Empathy* (New York: Random House, 2009).

"polarization," whose first order is physical. Let's look at what this means.

Simondon, like Jean Piaget later in *Biologie et connaissance* [Biology and Knowledge],[11] does not want to separate the thinking of the relationship between the living being and its milieu from a theory of knowledge, which he in fact seeks to rework so that knowledge is made into a complex form of adaptation of the living being understood as a "subject." More broadly still, we should be able to think what in *On the Mode of Existence of Technical Objects* he will refer to as "phases of culture"—technics, religion, art, science, etc.—as extending and complicating, through the play of *interlacements*, the tri-dimensional division of the living animal into "action," "perception," and "emotion." Thus, for example, "science is technical perception":[12] science and perception are both "psycho-somatic," adds Simondon, but the body of science is, one might say, technically decentered—while its psyche is *socially* decentered. So, this decentering, which is explicitly thought by Piaget, is what, for Simondon, "properly responds to a new engagement"[13] of the subject in the world: between perception and science there is *both* continuity and discontinuity.

This is why the living must be thought *on the one hand* as a *continual* individuation which, *on the other hand* and precisely because of this, holds in reserve the surprise of *its own overcoming*. So, what Simondon refers to as "individua*liza*tion" is simultaneously:

11 Jean Piaget, *Biologie et connaissance* (Paris: Gallimard, 1967). On the similarities as well as the differences between the approaches of Simondon and Piaget, see Victor Petit, "L'individuation du vivant (2). Génétique et ontogenèse," in *Cahiers Simondon no. 2* (Paris: L'Harmattan, 2010), 53–80.

12 Simondon, *L'individuation psychique et collective*, 140, and *L'individuation à la lumière des notions de forme et d'information*, 271.

13 Simondon, *L'individuation psychique et collective*, 140, and *L'individuation à la lumière des notions de forme et d'information*, 271.

- this permanent individuation of the living being, which is a "*theater* of individuation" and not only a "result of individuation" or of genesis;
- the *somato-psychic redoubling* of animal life, an ensemble of "sub-individuations" through which it becomes clear that "it is the psycho-somatic that is the model of the living being";[14] and
- what prepares, by creating the bio-psychic "subject," the conditions for psycho-social or "transindividual" individuation in which "personality" comes about.

Now, the strange idea, central to these three points, according to which the psycho-somatic is "the model of the living being," derives *only* from the Simondonian requirement that the living being be accounted for in its *becoming*—which takes it right up to the psycho-social—and it does not, therefore, lead Simondon to a vitalism that would cut the living being off from its pre-vital conditions. This is attested by the hypothesis of the distinct "orders" (the physical and the living) of the one *same* phenomenon of *polarization*: "we are in need of a systematic theory of polarization which would certainly further clarify the relations between what we call living matter (or organized matter) and inert or inorganic matter."[15] Simondon himself sketches out this theory of polarization, in the first place differentiating within the same phenomenon of polarization vital individuation from the individuation of the polarized crystal in formation:

> In the physical sphere, internal resonance characterizes the limit of the individual *individuating itself*; in the sphere of the living being, it becomes the criterion of the entire individual as individual; it exists in the individual's system and not only in the system that the individual forms with its milieu; the internal structure of the organism does not only result (as is

14   Simondon, *L'individuation psychique et collective*, 140, and *L'individuation à la lumière des notions de forme et d'information*, 271.
15   Simondon, *L'individuation et sa genèse physico-biologique*, 201, and *L'individuation à la lumière des notions de forme et d'information*, 203.

the case with crystal) from the activity taking place and the modulation happening at the limit between the spheres of interiority and exteriority; the physical individual, forever de-centered, forever peripheral to itself, active at the limit of its domain, has no true interiority; the living individual on the other hand does have a true interiority, because individuation happens on the inside; for the living individual, the interior is also constitutive, while for the physical individual only the limit is constitutive and what is topologically interior is genetically anterior. The living individual is contemporary with itself in each of its elements, which is not the case with the physical individual, which contains some past radically past, even when it is growing. Inside itself, the living being is an informational communication hub; it is a system in a system, comprising *in itself* the mediation between two orders of magnitude.[16]

Two comments on what Simondon says here:

Firstly, the difference indicated here between what Simondon will later call the "chrono-topology" of physical individuation (where what is "topologically interior" is "genetically anterior") and the chrono-topology of vital individuation (where the interior belongs to the present rather than the past), also coordinates with the Simondonian hypothesis of a topological—*which here is to say geometric*—peculiarity of the living being: "Nothing demonstrates to us that we could adequately think the living being through Euclidian relationships." This hypothesized geometric peculiarity of the living being is "topological" for Simondon *according to a non-Euclidian understanding of "topology"*: "living individuation must be thought according to topological schemas. Indeed, it is

16   Simondon, *L'individuation et sa genèse physico-biologique*, 26, and *L'individuation à la lumière des notions de forme et d'information*, 28.

by way of these topological structures that the spatial problems of the organism in evolution can be resolved."[17]

Secondly, attributing "true interiority" to the living individual is not the same as making it *substantial*—fighting against substantialism even being the whole point of Simondon's thinking of individuation.[18] Consequently, with respect to this interiority of the living being, Simondon clarifies:

> An immediate belief in the interiority of the being as individual comes, no doubt, from the intuition of one's body [*corps propre*] which seems, from the position of a thinking man, to be separated from the world by a material envelope which has a certain consistence and defines an enclosed space. In fact, a relatively deep psycho-biological analysis would show that, for a living being, the relation to the external environment is not distributed only at its external surface. The notion of the interior milieu, developed by Claude Bernard for the requirements of biological investigation, shows well enough through the mediation it establishes between the exterior milieu and the being, that the substantiality of the being should not be confused with its interiority, even in the case of the biological individual.[19]

The notion of polarization certainly represents Simondon's true Canguilhemian heritage,[20] and accordingly it responds to Canguilhem's fundamental *philosophical* interrogations: "In what is called a cell, it is biological individuality that is at issue. Is the individual a reality? An illusion? An ideal? No *one* science, not even biology, can answer this. And if *all* the sciences can and must make their contribution to this elucidation, it is doubt-

---

17   Simondon, *L'individuation et sa genèse physico-biologique*, 225, and *L'individuation à la lumière des notions de forme et d'information*, 227.

18   On this point, see my *Simondon ou l'Encyclopédisme génétique*, 9–19.

19   Simondon, *L'individuation et sa genèse physico-biologique*, 125, and *L'individuation à la lumière des notions de forme et d'information*, 127.

20   On this point we refer to Canguilhem's classic work, *Le normal et le pathologique* (Paris: PUF, 1966).

26 ful that the problem is properly scientific in the usual sense of the term."[21] After these words, Canguilhem adds the following remark as a note in the second edition of *La connaissance de la vie* [Knowledge of Life]: "Since these lines were written, Mr. Gilbert Simondon's thesis *L'individu et sa genèse physico-biologique* [The Individual and its Physico-biological genesis] (Paris: PUF, 1964) has thankfully contributed to the elucidation of these questions."[22] Indeed, as I have shown elsewhere in an extension of an article by Dominique Lecourt,[23] the question of knowing *where* individuality is situated—in the cell or the organism—*is no longer pertinent* for Simondon. This is because, from the inert molecule to the transindividualized personality, passing by cell and organism, we are in every case faced with increasing degrees of an individuality which is only ever a *result of individuation*:

> Strictly speaking, we cannot speak of the individual, but only of individuation; we must get back to the activity, to genesis, rather than trying to grasp the already given being in order to discover the criteria by which we can know whether or not it is an individual. The individual is not a being but an act, and being is an individual as the agent of this act of individuation by which it shows itself and exists. Individuality is an aspect of generation, is explained by the genesis of a being, and consists in the perpetuation of this genesis.[24]

---

21 Canguilhem, *La connaissance de la vie*, 2nd ed. (Paris: Vrin, 1969), 78 (author's emphasis). We find an echo of these words today in Alain Prochiantz's discourse on properly vital individuation: "[vital] individuation is a process without end, but also without purpose, whose comprehension draws on all fields of knowledge, including non-scientific disciplines, even if it falls to biologists alone to elucidate its mechanisms and conditions of existence" (*Machine-esprit* [Paris: Odile Jacob, 2001], 168–69).

22 Georges Canguilhem, *La connaissance de la vie*, 78.

23 See Dominique Lecourt, "La question de l'individu d'après George Canguilhem," in Bibliothèque du Collège international de philosophie, *Georges Canguilhem, philosophe, historien des sciences* (Paris: Albin Michel, 1993); see also my *Simondon ou l'Encyclopédisme génétique*, 17–19.

24 Simondon, *L'individuation et sa genèse physico-biologique*, 189, and *L'individuation à la lumière des notions de forme et d'information*, 191.

At which point we come back, *via* the concept of individuality, to **27**
individua*liza*tion, the first meaning of which was this "continuation of genesis," or continual individuation. We now understand
that, for each level of individuality, there is a corresponding level
of complexity of polarization: the polarization of the affectivity of
the bio-psychic animal "subject" is not the same as the polarization of the cellular membrane, which is not the same as the
undergoing of individuation of the crystal.

The Problem of Adaptation

The process of vital individuation described by Simondon will
bring him to criticize, even if allusively, what he calls the "biologism of adaptation." The criticism is primarily aimed at Darwin,
but Lamarck will also be targeted:

> Adaptation is correlative with individuation; it is only pos-
> sible in accordance with individuation. All biologism of
> adaptation, which is the basis for an important aspect of
> nineteenth-century philosophy, and which has come down
> to us in pragmatism, presupposes the already individuated
> living being as implicitly given; the processes of growth are
> partially left aside; it is a biologism without ontogenesis. The
> concept of adaptation in biology represents the projection
> of the relational schema of thought with an obscure zone
> between two clear terms, as in the hylomorphic schema;
> besides, the hylomorphic schema is itself present in the
> concept of adaptation: the living being finds forms in the
> world that structure the living being; the living being, on the
> other hand, gives form to the world so as to appropriate it:
> adaptation (passive and active) is understood as a recip-
> rocal and complex influence on the basis of the hylomorphic
> schema.[25]

25 Simondon, *L'individuation et sa genèse physico-biologique*, 207–8, and
    L'individuation à la lumière des notions de forme et d'information, 209–10.

28   Here it is Simondon's *anti-substantialism* that sustains the
critique. If life is an individuation or a perpetual genesis, then
"growth is not a separate process: it is the model for all vital
processes. . . . All functions of the living being are to some extent
ontogenetic, not only because they assure adaptation to an exter-
nal world, but because they participate in the permanent individuation of life."[26] The biological concept of adaptation is based
on a subtle and concealed substantialism which Simondon, with
reference to the great philosophical tradition deriving from Aris-
totle, calls "hylomorphism": "biologism of adaptation" is based on
the idea of an encounter between an *already given* individual and
an *already given* environment, each of which sometimes takes the
role of "form" and sometimes "matter." But nothing is given and,
moreover, genesis extends even *beyond* adaptation, as we see
with the living being that has become psycho-social, which rebels
rather than adapts itself.[27]

From which we understand that, once again, Simondon thinks of
life as a *becoming* by virtue of which "the psychic is, in this sense,
vital."[28] He summarizes his remarks with the following formula:
"individuation is anterior to adaptation, and is not exhausted in
it."[29] When, therefore, he insists on the fact that the biologism

26   Simondon, *L'individuation et sa genèse physico-biologique*, 207, and *L'individuation à la lumière des notions de forme et d'information*, 209.

27   In a "Supplementary Note" to *L'individuation à la lumière des notions de
forme et d'information*, Simondon aligns the difference between revolt and
adaptation and the difference between the living being and the machine—
which can adapt itself, but not revolt.

28   Simondon, *L'individuation et sa genèse physico-biologique*, 207, and *L'individuation à la lumière des notions de forme et d'information*, 209.

29   Simondon, *L'individuation et sa genèse physico-biologique*, 207 and *L'individuation à la lumière des notions de forme et d'information*, 209. It is possible to
speak here, with Alain Prochiantz on this occasion, of an "adaptation *through*
individuation" (my emphasis), which would "culminate with the human
brain and the invention of culture and language, which are unbelievable
instruments of individuation thanks to the significance they have for social
interactions in the construction of individuals" (Alain Prochiantz, *Machine-
esprit*, 166–67).

of adaptation is "a biologism without ontogenesis," he does not berate it for forgetting the conditions of adaptation that would be *less* than adaptation, but for reducing an activity of the living being that is *more* than adaptation to adaptation. Because *it is through actions and behavior that the living being develops*, and this activity which forms the individual instead of presupposing it is already *more* than adaptation.

Simondon clarifies a little later:

> In Lamarck, as in Darwin, we find the notion that the object is an object for the living being, an object that is constituted and detached, representing a danger, a foodstuff or a sanctuary. In the theory of evolution, the world in relation to which perception takes place is a world that is already structured according to a system of unitary and objective references. But it is precisely this objective conception of the milieu that distorts the concept of adaptation. There is not only an object as foodstuff or quarry, but a world defined by the search for food and a world defined by the avoidance of predators, or a world defined by sexuality. . . . The very concept of milieu is misleading: there is only a milieu for a living being which is able to integrate the perceptive worlds into a unity of actions. The sensory universe is not immediately given: there are only sensory worlds awaiting action in order to become significant. Adaptation creates the milieu and the being in relation to the milieu, the paths of the being; before action, there are no paths, no unified universe.[30]

Here Lamarck is also criticized, and we sense again, moreover, what will be explicitly confirmed later in the text: Simondon *purposely* mixes the two problematics of *adaptation* and *behavior*, because a thinking of life as becoming must be able *simultaneously* to think a radical genesis and an integration of complex

---

30   Simondon, *L'individuation et sa genèse physico-biologique*, 210–11, and *L'individuation à la lumière des notions de forme et d'information*, 212–13.

behaviors, to the extent that they still belong to the sphere of the living. The concept of adaptation that he criticizes as insufficient in order to think the living being designates a reaction *behavior,* where "passivity"—as *reaction*—is at the same time an activity with respect to the "adaptation" *that the theory of evolution refers to as* "fitness," *which does not relate to behavior.* Indeed, the cited passage continues with a critique of Kurt Lewin's psychology, since this psychology is based on the biological paradigm of adaptation. Elsewhere, Simondon relates embryogenesis to psychologist Arnold Gesell's "ontogenesis of behavior." We should note here that *the very importance of the paradigm of adaptation in the human sciences* has strengthened the reciprocal ambition in Simondon to think the living in such a way that complex behaviors can be accounted for.

This gesture is comparable in every respect to Erwin Schrödinger's in *Mind and Matter* (also in 1958), where he maintains that "Lamarckism is untenable," and at the same time rejects the "gloomy aspect of passivity apparently offered by Darwinism":[31]

> Without changing anything in the basic assumptions of Darwinism, we can see that the behavior of the individual, the way it makes use of its innate faculties, plays the most relevant part in evolution. . . . By possessing a new or changed character the individual may be caused to change its environment—either by actually transforming it, or by migration—or it may be caused to change its behavior towards its environment, all this in a fashion so as strongly to reinforce the usefulness of the new character and thus to speed up its further selective improvement in the same direction. . . . We must try to understand in a general way, and to formulate in a non-animistic fashion, how a chance-mutation, which gives the individual a certain advantage and favors its survival in a given environment, should tend to do

31   Erwin Schrödinger, *Mind and Matter,* in *What is Life?* (Cambridge: Cambridge University Press, 2014), 107.

more than that, namely to increase the opportunities for its being profitably made use of, so as to concentrate on itself, as it were, the selective influence of the environment.[32]

The perspectives put forward here, however, have no chance of shaking neo-Darwinian theory if they are not accompanied by an attempt to theorize *anew* that reality which we now know constitutes the fragile ground—fragile because it has not yet been thought in a sufficiently complex way—of molecular biology: the reality we call "information."[33] As long as this reality has not been properly reconsidered, the fragility of its current conception *will not be enough* to truly weaken neo-Darwinism. But, contrary to Schrödinger, whose work *What is Life*[34] was one of the sources for the informational paradigm of molecular biology as a reductionist theory of the "program," Simondon set about a timely and advanced critical interrogation of this Information Theory which,

32 Schrödinger, *Mind and Matter*, 107–10.

33 Among other things, the fact that the reality we call "information" is only applicable to the living being if it is rethought beyond the framework of its current theorization, was recalled by Michel Morange in a review article: "Some people have taken this notion of genetic information literally, tried to determine it quantitatively and to compare it to the quantity of information necessary for the creation of different living forms. This approach has a double weakness. The first is to imagine that genes, the genome, would by themselves be capable of allowing for the production of living organisms. . . . The second weakness of the notion of genetic information is that it describes badly the fundamental relationship connecting the sequence of nucleotides of DNA with the protein structure. . . . So we see how badly chosen the term information is for designating the role of genes and DNA, and how much better the term memory suits. . . . A second term taken from the field of information and used in biology also calls for analysis and criticism: the term program. Following François Jacob in *La Logique du vivant*, many biologists have used the term program to designate the action of genes in the development of living organisms. . . . This is to forget the hierarchical organization of the living being. Embryonic development can only be understood at the level of cells, tissues and organs. A uniquely genetic or molecular description of genetic development is impossible" ("Information," in *Dictionnaire d'histoire et philosophie des sciences*, ed. Dominique Lecourt [Paris: PUF, 2003], 526–27).

34 Schrödinger, *What is Life?* (Cambridge: Cambridge University Press, 2014).

a few years later, would come to sustain, *via* cybernetics and computer science, the "program" paradigm used in molecular biology. Of course, the creative power that may be *demanded* of such a critical questioning is not fully deployed by Simondon. But one may at least hope that it exists in his work as an as yet unrealized potential of his thought.

Information and Organization

We have seen that Simondon would like to think the living being as capable of integrating a psycho-social reality that cannot be reduced to the obscure laws of the survival of the species. In the same way, but without yet being able to interrogate the importation of an informational paradigm into biology—which had not yet taken place in 1958—he developed a critique of Information Theory as a *quantitative* theory specifically detached from the objective, which was unavoidable in his opinion, of accounting for *experiences of meaning*: as *experiences*, they are characteristic of the living being itself in its (inextricably *affective-perceptive-motor*) relations with its milieu. So, the *signification of* information is *both* what connects the living being to its psycho-social becoming, and what is left unthought by Information Theory.[35] Which brings Simondon to the following critique:

> Information theory is constructed to . . . allow a *correlation* between emitter and receiver in cases where this correlation has to exist; but if one plans to transpose it directly into the psychological and sociological spheres, it is paradoxical: *the narrower the correlation between emitter and receiver, the lower the quantity of information*. So, for example, in a fully completed apprenticeship, the operator needs only a very small quantity of information from the emitter, which is to say, from the object he is working on or the machine

35   On this last point, see Henri Atlan's now classic account in *L'organisation biologique et la théorie de l'information* (Paris: Hermann, 1972).

he is operating. The best form, therefore, would be that
which demands the lowest quantity of information. There is
something here that does not seem possible.[36]

After having made reference to Norbert Wiener[37] in order to take
up the new idea of thinking information as negentropic (an idea
proposed by Léon Brillouin as early as 1956),[38] Simondon here
declares his dissatisfaction. Ultimately, we would say that from
his perspective the purely *technical* objectives of Information
Theory tend, when the intention is to think the living being on
the basis of this theory, to produce a mysterious *break* between
the psychic and the biological, because Information Theory does
not seek to account for *signification*. Such, in any case, would be
Simondon's response to Henri Atlan's 1972 criticism of Olivier
Costa's desire to think, beyond the self-limitation of Information
Theory, "the enmeshing of psyche and matter."[39] Simondon too
would like *to be able* to think the living being as a psycho-somatic
relation, making use, for this purpose, of the *non*-technical
objectives of the Theory of Form—which is indeed a theory of

36   *L'individuation psychique et collective*, 51, and *L'individuation à la lumière des
     notions de forme et d'information*, 542 (author's emphasis). This text dates
     from 1960 and not 1958: it comes from the February 1960 conference of the
     Société Française de Philosophie, and, though it postdates it, was integrated
     into the published edition of Simondon's principal thesis. It is a better
     formulation of Simondon's refusal to reduce information to the probability
     schema of negentropy. For a technical analysis of this question, see my
     Penser l'individuation. Simondon et la philosophie de la nature, 116–30.

37   Norbert Wiener, *Cybernetics* (Cambridge, MA: MIT Press, 1948), and *The Use
     of Human Beings: Cybernetic and Society* (Boston: Da Capo Press, 1988). To
     a greater degree than Marxism, this second text is the true interlocutor of
     Du mode d'existence des objets techniques, while the first of these works by
     Wiener is only *one* of the major interlocutors of *L'individuation à la lumière
     des notions de forme et d'information*.

38   Léon Brillouin, *Science and Information Theory* (New York: Academic Press,
     1956).

39   On this point, see Atlan, *L'organisation biologique et la théorie de l'information*
     (Paris: Hermann, 1972), 196–200.

perception rather than of transmission, and which should apply even to the psycho-social.[40]

But *today*, Henri Atlan is in agreement with Simondon. Indeed, in *Le vivant post-génomique* [The Post-Genomic Living Being], after having remarked that "Simondon anticipated in this way the role of interference—'a certain margin of indeterminacy . . . which allows the machine to be sensitive to exterior information'—in both natural and artificial auto-organization,"[41] Atlan emphasizes the "inadequacies" of "Shannon's information theory": "on the one hand, its purely probabilistic nature which is seemingly ignorant of any question of signification, and on the other, the impossibility of information creation."[42] In decisively Simondonian style, Atlan then writes that "the 'genetic,' in the original sense of the term [i.e. *genesis*], is not in the 'gene.'"[43] The only remaining difference between Atlan and Simondon is that where the latter intends to *rethink* information, which had been too unilaterally probabilistic in Shannon, Atlan *subsumes* information as defined by Shannon into a more complex reality he calls "organization":

> There are implicit attributes in the idea of organization, which are opposed to each other in the way favored by the particular author. Indeed, on the one hand, we find complexity in the sense of unpredictability, variety, diversity, wealth of possibilities (of regulation and adaptation); the probabilistic function—Shannon's quantity of H information—may, in certain conditions, be a measure of this. But, on the hand, we

40 Reciprocally, Simondon rebukes the Theory of Form for not distinguishing the *whole* [*ensemble*] and the *system* [*système*], which is to say, for not thinking the *metastability* specific to the system. And this time he leans on Information Theory. On this game between the Theory of Form and Information Theory in Simondon, see my *Simondon ou l'Encyclopédisme génétique*, 70–71.

41 Atlan, *Le vivant post-génomique*, 24. Atlan's quotation of Simondon is taken from *Du mode d'existence des objets techniques*, 11.

42 Atlan, *Le vivant post-génomique*, 33.

43 Ibid., 55.

also find here attributes of order, regularity, repetition and internal constraints.[44]

This description by Atlan of the two aspects of *organization* echoes Simondon's description of the two aspects he saw in *information* itself:

> Information is, in one sense, something that can be infinitely varied, and something that requires, in order to be transmitted with minimal loss, that energy efficiency be sacrificed so as not to reduce in any way the range of possibilities. . . . But information, in another sense, is something that, in order to be transmitted, must be above the level of phenomena of pure chance, like white noise and thermal disturbance; so, information is something that has regularity, location, a defined sphere and a determined stereotypy by which it is distinguished from pure chance. . . . This opposition represents a technical antinomy which poses a problem for philosophical thought: information is like the chance event, and yet it is distinguished from it. An absolute stereotypy, excluding all novelty, also excludes all information. Yet, the distinction between information and interference is based on the reduction of the limits of indeterminacy.[45]

We see from this that what Simondon called "information," *as distinguished from Shannonian information*, corresponds with what Atlan calls "organization"—saying that it is irreducible to information... So, now we come to look at the way in which Simondon initiates a new theorization of information. Briefly put, his two major convictions, which will drive his effort to construct a *systemic and not cybernetic*[46] concept of information, are the following:

44 Ibid., 69–70.
45 Simondon, *Du mode d'existence des objets techniques*, 234–36.
46 On this distinction, see my *Simondon ou l'Encyclopédisme génétique*, 72–73. Even if thermodynamics, *via* the concepts of entropy and negentropy, has become a reference for thinkers of information, it should be recalled that

- The fundamental condition for there to be information is not a particular state of the emitter, nor is it a property of the message, but a particular state of the *receiver*, which Simondon qualifies as "metastable" because it is charged with potentiality so as to make *becoming*-informed possible.
- This information as the transmission of the message is nothing but *a perpetuated genesis* of the receiver—because *all* information is genesis—and there is a "first information" in which emitter and receiver *do not yet exist*. The condition of possibility here is a first metastability which is picked up by the information receiver when information is message transmission.

Because of these two convictions, which make message transmission a *particular instance* of information, it is a matter, for Simondon, of thinking a universal process of information, with this latter in fact being the "formula of individuation."[47] From the formation of a crystal to the signification experienced by the transindividuated personality, and by way of genetic information and organic perception, we are dealing with different "phases" of the same process of information, understood as genesis or individuation, with these different phases able to coexist in a *multi*-phased individual. But one last hypothesis, explicitly presented as such by Simondon, but from early on and repeatedly—making it in some respects foundational—says that vital individuation is only the continuation of an initial inchoate phase of physical individuation. In other words, the relation between vital individuation

information theoreticians and cyberneticians (as suggested by Bertalanffy, so as to distinguish himself from them, in his *General System Theory* [New York: George Braziller, 1968]) did not in the first place draw on thermodynamics, which on the contrary inspired systems theory—which did not however make use of the idea of entropy, but of "metastability" (Simondon) as "a dynamic interaction of components" (Bertalanffy). Those we refer to as theoreticians of "complexity," like Simondon in philosophy or Henri Atlan in science, are in this way closer to systems theory than to cybernetics, whose paradigms are essentially to be found in technology rather than contemporary physics.

47   Simondon, *L'individuation et sa genèse physico-biologique*, 29, and *L'individuation à la lumière des notions de forme et d'information*, 31.

and physical individuation would be a *neotenic* relation. The interest of this hypothesis is certainly that, for all that, it enables the
radical thinking of genesis to avoid falling back into a reduction
of the living to the physical: here, the living, being individuated
like the physical, has its origin in a *"pre*-individual" reality which is
qualified by Simondon as "pre-physical and pre-vital."

Apoptosis and Permanent Ontogenesis

What, in conclusion, are the possible links between Simondon's
perspectives and Jean-Claude Ameisen's work on apoptosis? Let
us recall first of all that in *La sculpture du vivant* [The Sculpt*ing
of* the Living Being] Ameisen argues that apoptosis or "cellular
suicide" *participates* in the ontogenetic process itself. In *Simondon
ou L'Encyclopédisme génétique*, I believed it possible to say that
a *first* link between *La sculpture du vivant* and *L'individuation à la
lumière des notions de forme et d'information* could be found in this
idea of death's constitutive role in life itself, because Simondon
had distinguished between death which "translates the very
instability of individuation, *its confrontation with the conditions of
the world*," and death which *"does not come from the confrontation
with the world*, but from the convergence of internal transformations."[48] He clarified:

> for the living being, death exists in two forms which do not
> coincide: it is adverse death . . . . But death also exists for
> the individual in another sense: the individual is not pure
> interiority: it grows heavy with the residual weight of its
> operations; it is passive in itself; it is its own exteriority . . . .
> In this sense, it seems that the fact that the individual is not
> eternal need not be considered accidental; the whole of life
> can be considered as a transductive series; death as the
> final event is only the consummation of a deadening process

48   Simondon, *L'individuation et sa genèse physico-biologique*, 213, and *L'individuation à la lumière des notions de forme et d'information*, 215 (my emphasis).

that accompanies every vital operation as an operation of individuation; each operation of individuation leaves death in the individuated being which is progressively loaded with something that it cannot eliminate; this deadening differs from the degradation of organs; it is essential to the activity of individuation.[49]

In my brief commentary of this passage, I added that, without seeing here a strict anticipation of the thesis of apoptosis as the very condition of life, one should at least recognize that Simondon *integrates* death into the process of life as permanent individuation. I would like to clarify here both *the meaning and the limits* of this possible parallel between Simondon's hypothetical speculations and the most recent advances of cellular biology and immunology. It will appear that *even if Simondon does not think apoptosis, there is at least the intuition of a new theory of aging according to which the latter is not only wear and tear, but also points to the constitutive role of death for life, which is thought today by Ameisen* via *the link between "death before the fact" and reproduction.* This intuition of Simondon's on the subject of aging is in evidence in this statement from the center of the cited passage: "death as the final event is only the consummation of a deadening process that accompanies every vital operation as an operation of individuation." But this is only an intuition with all its inherent limitations, which will come to light in a brief analysis. Let us see how things stand.

The distinction between death as *terminus* and death as an *internal condition* is explicit in Simondon's passage in the distinction between "wear and tear" and "deadening," a distinction which, at first sight, is that much more obscure since Simondon explains the aging phenomenon by way of the second, and it is difficult to understand—still at first sight—how it would not be based on the first. Only the notion of a "sediment," articulated here by

49   Simondon, *L'individuation et sa genèse physico-biologique*, 213, and *L'individuation à la lumière des notions de forme et d'information*, 215.

Simondon *via* the words "each operation of individuation *leaves* death in the individuated being", allows us to distinguish at a push between wear and tear and deadening, but this notion is not *on its own* what will truly enable us to make death into a process conditioning life: in order for the notion of a sediment to itself contribute to making deadening something "essential to the activity of individuation," as Simondon says when he distinguishes it from wear and tear, the sediment cannot be a simple sediment. Now, it is precisely with respect to the phenomenon of aging that Ameisen allows Simondon's hypothetical and still simply intuitive speculations to take on their full meaning—through an extension/overturning of the notion of a sediment:

> Aging and death may not only result from wear and tear, from the passage of time and the body's inability to withstand the assaults of the environment. . . . A protein [issuing from the *Methuselah* gene], a minimum production of which is essential to the construction of the embryonic body [of the fruit fly], also has the effect of shortening the lifespan of adults when it is produced—beyond this minimum threshold—in a 'normal,' which is to say, excessive quantity. A minimal production of the Methuselah protein favors individual longevity, but risks compromising fecundity; an excessive production favors premature aging but brings a margin of security to the propagation of the species.[50]

Simondon had made reproduction "pre-eminent amongst transductions," which is to say, the radical form of vital individuation, and he had also intuited that aging does not proceed only from wear and tear. But, because he did not have at his disposal this notion of an *intrinsic constitutive role of death for life* provided by the new theorization of apoptosis, he did not *bind* reproduction to death except in the classic and so to speak metaphoric form of the extension of self to the after-self. It remains the case that, to the same extent that apoptosis *properly speaking* is not at stake

50   Jean-Claude Ameisen, *La Sculpture du vivant* (Paris: Seuil, 2003), 374 and 384.

in the new theory of aging put forward by Ameisen, it is possible to maintain that Simondon, at the level of the intuitions that motivated him, would have been perfectly in accord with these words from the biologist, dedicated this time to "splitting" and *cellular* aging:

> Each time that the mother-cell splits its genetic library before generating, it also splits, on the basis of its chromosomes, little supernumerary copies of circular DNA. And it keeps these copies, which are not allotted to the daughter-cell, in itself. As the mother-cell continues to give birth, its body contains an ever increasing number of copies. The accumulation of these little DNA circles above a certain threshold seems to trigger the fragmentation of the nucleus of the mother-cell and its death. . . . The idea is that life's victory over wear and tear is bound to a local heightening of disorganization—of the advance towards disorder—in one part (the mother-cell) which enables the birth in another part (the daughter-cell), of a discrete, local level of order and complexity. The passing of a maternal body is accelerated to enable the birth and survival of an infant body.[51]

So we come in conclusion to the second of the two links between Simondon and Ameisen. That is to say, to the idea, introduced by Ameisen at the end of *La sculpture du vivant*, according to which

> dramatic changes in the environment can bring to light, in a body that is developing itself, a pre-existent source of novelty—a potentiality—which had accumulated progressively over time and which, continually repressed until now, is suddenly able to show itself for the first time. In this way, the external environment has the power to sculpt the living being.[52]

I will give two successive readings of these words:

51   Ibid., 416–17.
52   Ibid., 409.

a. Even if Ameisen does not say as much—we will see *why* in the
second reading—the process he describes corresponds, at least
in the first instance, to what Stephan Jay Gould called "exaptation,"
and which Pierre Sonigo, in a commentary on the latter, distinguished from the idea of "programmed anticipation" suggested
by the Darwinian term "pre-adaptation": "evolutionary innovation is brought about by unexpected encounters between the
potential and the useful."[53]

Whatever the case may be, the schema proposed by Ameisen of
a revelation, through the agency of the dramatically altered environment, of a potentiality that has been progressively accumulated in the organism, aims explicitly to mediate, and ultimately
go beyond (in agreement, I would add, with Gould's point of view),
the opposition between the "gradualists" and the "punctualists,"
which is to say, between the conception of evolution in terms of
negligible genetic modifications and the conception of evolution
in terms of "leaps ahead" or sudden jumps. This mediation and
this move beyond oppositions would have appealed to Simondon
who himself sought systematically to subvert naïve alternatives
and who, in *Du mode d'existence des objets techniques*, had in the
same way associated continuity and discontinuity in order to
think the becoming of the technical object. In the thinking of the
living being, the way in which Ameisen pays particular attention
to the question of as yet unrealized potential seems to be truly
Simondonian, as does the possibility of formulating anew a concept on which Simondon was particularly reliant: the Bernardian
concept of "interior milieu," which indeed Ameisen seems to
revisit and which is not antithetical to his thesis according to
which "the environment is more than a simple filter—a bot-
tleneck—through which individuals and species are selected or
eliminated. The exterior environment can exert a direct influence

53   Pierre Sonigo and Isabelle Stengers, *L'évolution* (Paris: EPD Sciences, 2003),
     53.

on the way in which cells and bodies use their genetic potentialities and so on the manner in which embryos are constructed."[54]

b. The way in which Ameisen establishes this last idea would have further appealed to Simondon since the biologist intends to differentiate himself from Gould here (who, moreover, Ameisen places on the side of the punctualists), to the extent that for Gould, as for the (opposed) positions of gradualism and punctualism, the "emergence of individuals and species endowed with new properties is considered to be an immediate translation, a direct consequence in real time, of the appearance of chance modifications in their genes . . . . In other words, the essential debate between these two theories does not concern the way in which the environment sculpts the new aspect, but the nature of the modifications on which it brings its effects to bear."[55] In opposition to this common point of view, which had underpinned the debate up to now, Ameisen suggests developing the consequences of the work of Linquist and Rutherford, which he expounds as follows:

> When the embryos of fruit flies undergo a thermal shock, the new-borns exhibit profound modifications in a whole range of organs—antennae, wings, eyes, legs. These modifications vary from one embryo to another and from one sub-species of fruit fly to another. The appearance of this new aspect is not connected to the sudden appearance of genetic modifications: it is due to the revelation of a pre-existent genetic diversity, whose appearance had until now been permanently repressed.[56]

This "repression" is performed by proteins which are called "chaperones" and which attach themselves to modified proteins, "allowing them to return to their initial form."[57] And so

54   Ameisen, *La sculpture du vivant*, 406.
55   Ibid., 405–6.
56   Ibid., 408.
57   Ibid., 407.

it is necessary to give a *second reading* of the above citation
from Ameisen in which it was said that "dramatic changes in
the environment can bring to light, in a body that is developing
itself, a pre-existent source of novelty—a potentiality—which
had accumulated progressively over time and which, *continually
repressed until now*, is suddenly able to show itself for the first
time." The words I have italicized contain the theoretical innovation which means that it is no longer necessary for Ameisen to
refer to Gould's exaptation: here, the action of the new environment, which suddenly reveals potentialities accumulated in the
organism, no longer operates on the form of the proteins—so on
the gene's mode of expression—but on the internal agents which
until now restored this form when it had been altered in this
way. This is the "complexity" explicitly claimed by Ameisen, and
"complexity" is the watchword whose great pioneer, as I showed
in *Simondon ou l'Encyclopédisme génétique*, is without doubt
Simondon. So we have our work cut out for us.

Works Cited

Ameisen, Jean-Claude. *La Sculpture du vivant*. Paris: Seuil, 2003.

Atlan, Henri. *L'organisation biologique et la théorie de l'information*. Paris: Hermann, 1972.

———. *Le vivant post-génomique, ou Qu'est-ce que l'auto-organisation?* Paris: Odile Jacob, 2011.

Barthélémy, Jean-Hugues. *Penser l'individuation: Simondon et la philosophie de la nature*, Paris: L'Hartmattan, 2005.

———. *Simondon ou L'Encyclopédisme génétique*. Paris: PUF, 2008.

Bertalanffy, Ludwig von. *General System Theory*. New York: George Braziller, 1968.

Brillouin, Léon. *Science and Information Theory*. New York: Academic Press, 1956.

Canguilhem, Georges. *La connaissance de la vie*. 2nd ed. Paris: Vrin, 1969.

———. *Le normal et le pathologique*. 2nd ed. Paris: PUF, 1966.

Dagognet, François. Preface to Bernard, Claude. *Introduction à l'étude de la médecine expérimentale*, 9–21. Paris: Flammaron, 1984.

Fagot-Largeault, Anne. "L'Individuation en biologie." In Bibliothèque du Collège international de philosophie, *Gilbert Simondon: Une pensée de l'individuation et de la technique*, 19–54. Paris: Albin Michel, 1994.

Gayon, Jean. *Darwin et l'après-Darwin*. Paris: Kimé, 1992.

44   Lecourt, Dominique. "La question de l'individu d'après George Canguilhem." In *Bibliothèque du Collège international de philosophie, Georges Canguilhem: Philosophe, historien des sciences.* Paris: Albin Michel, 1993.

Leroi-Gourhan, André. *Le geste et la parole.* Vol. 1 and 2. Paris: Albin Michel, 1964–65.

Morange, Michel. "Information." In *Dictionnaire d'histoire et philosophie des sciences,* edited by Dominique Lecourt, 526–27. Paris: PUF, 2003.

Petit, Victor. "L'individuation du vivant (2). Génétique et ontogenèse." *Cahiers Simondon no. 2,* edited by Jean-Hugues Barthélémy, 53–80. Paris: L'Harmattan, 2010.

Piaget, Jean. *Biologie et connaissance.* Paris: Gallimard, 1967.

Prochiantz, Alain. *Machine-esprit.* Paris: Odile Jacob, 2001.

Schrödinger, Erwin. *Mind and Matter.* In *What is Life?* Cambridge: Cambridge University Press, 2014.

Simondon, Gilbert. *Du mode d'existence des objets techniques.* Paris: Aubier, 1958.

———. *L'individu et sa genèse physico-biologique.* Grenoble: Millon, 1995.

———. *L'individuation à la lumière des notions de forme et d'information,* Grenoble: Millon, 2005.

———. *L'individuation psychique et collective.* Paris: Aubier, 1989.

Sonigo, Pierre, and Isabelle Stengers, *L'évolution.* Paris: EPD Sciences, 2003.

Stiegler, Bernard. *La Technique et le Temps.* 3 vols. Paris: Galilée, 1994–2001.

Tibon-Cornillot, Michel. "Penser en amont de la bioéthique: transformations dirigées du génome et crise du néodarwinisme." In *Vers un anti-destin? Patrimoine génétique et droits de l'humanité,* edited by François Gros and Gérard Huber, 127–46. Paris: Editions Odile Jacob, 1992.

Waal, Frans de. *The Age of Empathy.* New York: Random House, 2009.

Wiener, Norbert. *Cybernetics.* Cambridge, MA: MIT Press, 1948.

———. *The Use of Human Beings: Cybernetic and Society.* Boston, MA: Da Capo Press, 1988.

Technology and the Question of Non-Anthropology

Introduction: Non-Anthropology; or, The Conditions of a Dialogue

What are the initial conditions of a dialogue between Simondon and Heidegger? If the question arises, it is because the difference between these two thinkers at first seems irreducible, to such a degree that the dialogue risks, through an absence of common ground, becoming a misunderstanding. But common ground there is, and it is most apparent with respect to the major theme of technology.[1] To be precise, it is not a misunderstanding that will underwrite the dialogue between Heidegger and Simondon, but their mutual demand for a non-*"anthropological"* thinking of technology.

---

1   Translator's note: I have translated both *Technik* (from German) and *technique* (from French) as "technology" for the sake of consistency with the existing translations of Heidegger's essay "The Question Concerning Technology"; it is also noted that Simondon uses the French word "technologie" when he refers to the study of "technique(s)."

For Simondon, the word "anthropology" is not to be understood in any of its classical senses but designates, on the one
hand, an essentialist thinking which cuts human being off from
the living, and on the other, a thinking that reduces technology
to its use by human being—to what Simondon refers to as the
"labor paradigm." It is Simondon's critique of this second aspect
of anthropological thought that we will soon be examining. For
the moment we emphasize that whereas Simondon's critique of
these two aspects of anthropology has led some to suppose that
he was an anti-humanist, his stance seeks only to rehabilitate
technology, on the one hand, and philosophy of nature on the
other, without this being to the detriment of human being. It is
only really possible to understand Simondon here if we accept
that he privileges a subversion of classical conceptual oppositions
and conflicts of views.[2] So, in both *Du mode d'existence des objets
techniques* [On the Mode of Existence of Technical Objects] (which
was his supplementary thesis for his *doctorat d'État*) and *L'individuation à la lumière des notions de forme et d'information* [Individ-
uation in the Light of Notions of Form and Information] (which
was his main thesis), the French philosopher rejects the conceptual oppositions of nature/culture and technology/culture. It is
the presence of a third classical opposition—between nature and
technology—that betrays the weakness of these two conceptual
oppositions. It is precisely by subverting this third opposition that
Simondon will subvert the first two.

2   I dedicated the whole of Chapter II of my book *Simondon* (Paris: Belles
     Lettres, 2014) to this privileging. Here I will simply recall that Simondon not
     only rejects the alternatives realism/idealism, empiricism/innatism and
     skepticism/dogmatism, which had already been challenged by Kant and his
     successors, but also the oppositions mechanism/vitalism, psychologism/
     sociologism and humanism/technicism, against which he directed the bulk
     of his work. I showed that underlying all these theoretical alternatives
     there is the conceptual opposition matter/form, which is hidden in each of
     the theses. It is to this opposition that Simondon, for his part, brings back
     the opposition between the philosophizing subject and his object that had
     already been interrogated by Heidegger.

Simondon's privileging of a subversion of classical oppositions was rightly highlighted by Gilles Châtelet in his article "Simondon" for the *Encyclopaedia Universalis*, and with this privileging Simondon is closer to the book *Pour l'homme* [For Man] by his phenomenologist friend Mikel Dufrenne than to so-called anti-humanist thinkers. Indeed, *Du mode d'existence des objets techniques* criticizes what it calls "an easy humanism,"[3] which is not humanism in general, but only what one might call "a far too easy humanism." In opposition to this, it is for him a matter of establishing what I would like to call a "difficult humanism,"[4] which is to say, a humanism compatible with the critique of the two aspects of "anthropology" as defined by Simondon. On the one hand, this difficult humanism integrates human reality into *physis*, and on the other technology into culture. These two integrations are in fact for Simondon just one, since technology, he says, is itself what "expresses" "nature" in its connection with the "subject": the technical object is the extension of life through which that life can go beyond itself in a relationship referred to as "transindividual." So, Simondon says, the technical object is nature having become a "support" for what extends and overcomes simple life. This, briefly stated, is the subversion of the first two classical oppositions by way of the subversion of the third.

Now, for Heidegger, the word "anthropology" once again refers to a twofold naivety: on the one hand, anthropology is a thinking that reduces the essence of human being to a "present-at-hand being" (*étant-là-devant / vorhandenes Seiende*), when this essence is, for Heidegger, "Being-there" (*Dasein*); on the other hand, anthropology is a thinking that reduces technology to its use by human being. This second aspect brings us back to what is also

3   Gilbert Simondon, *Du mode d'existence des objets techniques* (Paris: Aubier, 1958), 9.

4   I have developed this theme in two texts: "What New Humanism Today?," trans. Chris Turner, *Cultural Politics* 6, no. 2 (2010): 237–52; and "L'humanisme ne prend sens que comme combat contre un type d'aliénation" (interview with Ludovic Duhem), *Tête-à-tête*, no. 5 (2013): 54–67.

rejected by Simondon. So, it is here that a true dialogue is possible, and the question that we will pose to set this dialogue up is, of course, the following: How, in its concepts and its theses, should a non-anthropological thinking of technology take shape? Now, the common thesis of non-anthropology will not be understood in the same way by the two thinkers, not only with respect to the first meaning of the word "anthropology," but also with respect to the second. This is the situation: the first aspect of anthropology as defined by Heidegger does not match the first aspect of anthropology as defined by Simondon. Saying that anthropology reduces the essence of human being to a present-at-hand being is not the same as saying that anthropology cuts human being off from the living. Each of the two thinkers would situate the other within anthropology—for Simondon, Heidegger is still effecting an essentialist break while, for Heidegger, Simondon is still effecting a reduction of the essence of human being to present-at-hand being.[5] But for Simondon, as for Heidegger, the two aspects of the anthropology they set out to challenge go together, so the understanding of the second aspect will also differ between the two, despite the verbal similarity possible in the initial diagnosis. The construction of non-anthropology will therefore be different for each of our two thinkers, as much in its second as in its first aspect. This is what we must now confirm by bringing our analysis to bear upon only the second aspect of non-anthropology—the aspect relating to technology—and by beginning with an exposition of Simondon's position.

---

5   It *doesn't* mean that Simondon reduces human being to a thing. What Heidegger criticizes as an "objectification" (*Ver-gegen-ständlichung*) is not what in French is called *"chosification"* (*Verdinglichung*) but *a more general attitude of knowledge* that one might call "objectivation." I have authorized the translator to use the *classical* English translation, that is to say "objectification," but one must keep in mind these remarks.

The Non-Anthropological Thinking of Technology in Simondon

L'individuation à la lumière des notions de forme et d'information considered the individuation—which is to say, in Simondon's work, the *genesis*—of physical, vital, and psycho-social or "transindividual" beings. *Du mode d'existence des objets techniques* considers the individuation of technical beings in that they are also a genesis. It is only by way of the latter that it is possible, according to Simondon, to bring out the sense of technical objects, and to reinstate technics with respect to its participation in culture. This is why the first part of the book is titled "Genèse et évolution des objets techniques" ["Genesis and Evolution of Technical Objects"]. From the start, it is a question of refusing to define the technical object starting from a classification into genres and species of the individual considered as a given. Here, as in his main thesis, "it is better to reverse the problem: it is starting from criteria of genesis that it is possible to define the individuality and specificity of the technical object."[6] But it is the *labor* paradigm which, for Simondon, seems to order the traditional classification of technical objects into genres and species. This labor paradigm, when it is considered as a *social relationship* between a master and a slave, is an *unconscious* paradigm of the hylomorphism against which Simondon is fighting in his work, the *conscious* paradigm of which lies, he says, in the *technical* operation of casting bricks.[7] Indeed, it is the labor paradigm that orders the reduction of technical objects to their usage, which in turn defines the genres and species whose illusory character is condemned by Simondon:

---

6   Simondon, *Du mode d'existence des objets techniques*, 20.
7   See Simondon, *L'individuation à la lumière des notions de forme et d'information* (Grenoble: Millon, 2005), Part I, Chapter I; and my commentary in *Penser l'individuation: Simondon et la philosophie de la nature* (Paris: L'Harmattan, 2005), Chapter II, 4. This commentary is taken up again in abridged form in my book *Simondon*, 106–110.

species are easy to summarily distinguish, for practical use, as long as it is accepted that the technical object is to be understood with respect to the practical end to which it responds; but this is an illusory specificity because no structure corresponds to a defined use. The same result can be obtained starting from very different operations and structures: a steam engine, a petrol engine, a turbine, and spring or weight-powered engines are all engines; but the spring-powered engine is in fact more closely analogous to a bow or a crossbow than to a steam engine; the motor of a weight-driven clock is analogous to a winch, while an electric clock is analogous to a doorbell or buzzer. Use brings heterogeneous operations and structures together in the same genres and species which take their signification from the relationship between this operation and another, that of the human being in action. So, what we call by a single name—engine, for example—may be multiple at a given moment and may vary with time, changing character.[8]

Here it is the opposition between utilitarian character and operation that is central. That the classification of technical objects into genres and species according to their use derives from the unconscious paradigm of hylomorphism constituted by labor, is apparent in this passage, even if implicitly and allusively, in the notion of the subsumption of the object under its use by "the human being in action." Such is the root of what Simondon regards as an "anthropological" reduction of technology. The different engines, for example, only have "a single name" thanks to this illusory subsumption of operation under use, through which the only thing that can truly define a technical object is lost—its genesis:

8   Simondon, *Du mode d'existence des objets techniques*, 19. [The word "operation" is here used to translate *"fonctionnement,"* because of the ambiguity of "functioning" and "working": the first doesn't make a clear distinction between the operation and the function or use, and the second refers to work, which Simondon rejects as a blinding paradigm. —Auth. & Trans.]

The unity of the technical object, its character and specificity, **53** are characteristic of the consistency and convergence of its genesis. The genesis of the technical object appertains to its being. . . . The petrol engine is not just any such engine in time and space, but the fact that there is a development, a continuity from the first engines to those that we know, which are still evolving. In this respect, as in a phylogenetic lineage, a definite evolutionary stage contains in itself dynamic structures and schemas which underlie an evolution of forms. The technical being evolves by convergence and by self-adaptation; it inwardly coheres according to a principle of internal resonance.[9]

Both the beginning and the end of this passage show how the technical object is considered here in terms of the perfecting (*perfectionnement*) of a pre-existing operation describing a "lineage." So the opposition is not between genesis and progress, but between progress with respect to operation and progress with respect to usage, with the latter conforming to totally different criteria than those defining the progress of operation as the genesis of the technical object: "for this or that use, an engine from 1910 is superior to an engine from 1956."[10] Indeed, true technical progress conforms to a principle of "convergence" and unification by virtue of which a reciprocal causality is established through which each element receives its form:

In a contemporary engine, every important element is so bound up with the others through reciprocal exchanges of energy that it cannot be other than it is. The form of the combustion chamber, the form of the valves and the form of the piston belong to the same system in which there are a multitude of reciprocal causalities. . . . One could say that the contemporary engine is a concrete engine, while the former engine is an abstract engine. In the former engine, each

9   Ibid., 20.
10   Ibid.

element takes part at a given moment in the cycle, and then it is no longer supposed to act on the other elements.[11]

This is what Simondon, here taking up the Hegelian notions of the abstract and the concrete, calls the process of "concretization" of technical objects. But such a reciprocal causality only "has" its "truth"—again in Hegelian terms—in the idea of the poly-functionality of elements, which alone allows the process of "concretization" to be defined as a process of "convergence." Contrary to the motives determining the usage of a technical object, however, the motives determining its evolution by convergence are not strictly speaking anthropological: "if technical objects evolve in the direction of a small number of specific types, this is due to an internal necessity and does not depend on economic influences or practical requirements."[12] But before clarifying how the "non-anthropological" character of the process of concretization–convergence should be understood, we should note that Simondon, if we follow the text, must consider the *bespoke* objects of the artisan as "abstract," in opposition to industrial objects which alone are "concrete": "at the industrial level, the object has acquired its coherence, and it is the system of needs that is less coherent than the system of the object; needs mold themselves to the industrial technical object, which in this way acquires the power to fashion a civilization."[13] These last words point to the progressive autonomization of the process of concretization-convergence, whose "internal necessity" is asserted by Simondon:

> The structural reforms allowing for the specification of the technical object constitute what is essential to the becoming of that object; even if the sciences make no advance during a given period, the progress of the technical object towards specificity can continue to take place; indeed, the principle of this progress is the way in which the object brings itself

11   Ibid., 21.
12   Ibid., 23–24.
13   Ibid., 24 (author's emphasis).

about and conditions itself in its operation and in the
reactions of its operation on use; the technical object, issuing
from an abstract labor organizing sub-ensembles, is the
scene of a certain number of relations of reciprocal causality.
It is these relations which mean that, based on certain limits
in the conditions of utilization, the object discovers obstacles
within its functioning: *it is in incompatibilities produced from
the progressive saturation of the system of sub-ensembles that
we find the play of limits whose overcoming is constitutive of
progress*; but it is in its nature that this overcoming can only
take place through a leap, through a modification of the
internal distribution of functions, a rearrangement of their
system; what was an obstacle must become a means of
realization.[14]

The end of this passage brings us back to what Simondon calls,
in a note from the same passage, the "conditions of individuation of a system," conditions which mean that "the specific
evolution of technical objects is not completely continuous, nor
is it completely discontinuous."[15] Because technical progress in
fact changes the obstacles themselves into solutions, it happens
by continuous supersaturation and discontinuous individuation,
with supersaturation being found in incompatibilities balanced
by "detail refinement" of a structure which they do not reorganize
but which they end up revealing as problematic, the new individuation being the solution which uses the incompatibilities—
simultaneously balanced and revealed by these adjustments—to
reorganize the structure itself.

Now that we have made these remarks, we can come back to
what I referred to as the "non-anthropological" character of the
process of concretization–convergence. Simondon distinguished
between the intention on which the fabrication of a technical
object is based, which is connected to its operation, and the

14   Ibid., 27–28 (author's emphasis).
15   Ibid., 27.

intention on which its use is based. But the fabricating intention
can only explain the genesis of the technical object on condition
that this intention is not considered anthropologically, which is
to say as originating with a meaning-giving subject similar to the
user. In this sense, Simondon does not oppose the Heideggerian
thinking of *Gestell*: neither of these two thoughts is—at least at
first sight—anthropological, even if Heidegger does not situate
the non-anthropological thinking of technology within fabrication. Connected with this restriction there is, as we shall see,
a real incompatibility in another sense between these two great
thinkers of technology, since Simondon would certainly not have
agreed with the Heideggerian thesis according to which "the
essence of technology is nothing technological."[16] It is a question
of knowing whether this thesis, which "ontologizes" technology in
order to "deanthropologize" it, does not originate in an anthropological blind-spot with respect to fabrication, at least from a
Simondonian perspective, if the expression "nothing technological" closing the Heideggerian formula means "nothing of a human
operation or means." We will now undertake an internal critique
of the Heideggerian thinking of technology, which is to say, a
critique setting out from the non-anthropological intention of
this thinking so as to turn this intention against the Heideggerian
mode of its realization.

The Non-Anthropological Thinking of Tech-
nology in Heidegger: Towards an Internal
Critique of *Gestell*

The way in which Heidegger quite rightly challenges the
anthropological thinking of technology may, if examined from
Simondon's point of view, in fact seem still metaphysical, even

16   Martin Heidegger, "The Question Concerning Technology," in *The Question
      Concerning Technology and Other Essays*, trans. William Lovitt (New York:
      Harper Perennial, 1977), 4. [Translation slightly modified. —Trans.]

"anthropological" in the profound Simondonian sense of the term: for Heidegger, it is still in the name of the essence of human being that the essence of technology is said to have "nothing technological" in itself, which is to say nothing of a simple human operation or simple human means. It is surely not by chance if ultimately "the essence of technology cannot be guided into the metamorphosis of its fate without the aid of human being."[17]

We recall in this connection the major steps of "The Question Concerning Technology."[18] In the same way that Simondon had distinguished between technology as operation, on the one hand, and the use to which we habitually reduce it on the other, Heidegger distinguishes between the "essence of technology" and its common representation as a means directed towards an end.[19] Even if this is "correct,"[20] both with respect to the technology of the artisan and to modern technology, the anthropological conception of technology misses, for Heidegger, the true—and no longer simply "correct"—essence of technology. It is an essence that, on this occasion, only modern technology leads us to question: "it is precisely the latter [modern technology] and it alone that is the disturbing thing, that moves us to ask the question concerning technology per se."[21]

17   Martin Heidegger, "The Turning" in *The Question Concerning Technology and Other Essays*, 39. [NB. The English translation differs in a number of ways from the French version. Here I have translated from the French. The unaltered English version reads as follows: "the coming to presence of technology cannot be led into the change of its destining without the cooperation of the coming to presence of man." —Trans.]

18   For a commentary, see Jacques Taminiaux, "L'essence vraie de la technique," in *Cahier de l'Herne: Heidegger*, ed. Michel Haar (Paris: Editions de l'Herne, 1983), 263–84.

19   As I said earlier, Heidegger nevertheless situates fabrication on the side of use, and so in an "instrumental and anthropological definition of technology" ("The Question Concerning Technology," 10). We may conclude from this that Simondon remains naive, but it is in fact Heidegger who shifts the non-anthropological nature of the process of concretization in favor of a *destinal* thinking of technology, as we shall see.

20   Heidegger, "The Question Concerning Technology," 5.

21   Ibid., 13–14.

One must, nevertheless, distinguish between the essence of
modern technology and the essence of technology: even if one
leads to the questioning of the other, it cannot be reduced to it,
and it is precisely for this reason that the questioning it brings
about is "disturbing." To start with, the essence of technological
bringing-forth (*pro-duction*) is not artificial fabrication but the
"disclosure" from which bringing-forth proceeds, which is *physis*
itself:

> It is of the utmost importance that we think the bringing-
> forth [*la pro-duction, Hervor-bringen*] in its full scope and at
> the same time in the sense in which the Greeks thought it.
> . . . *Physis* also, the arising of something from out of itself, is
> a bringing-forth [*une pro-duction*], *poiēsis.* . . . Occasioning
> has to do with the presencing [*Anwesen*] of that which at any
> given time comes to appearance in bringing-forth. Bringing-
> forth [*Le pro-duire*] brings hither out of concealment forth
> into unconcealment. Bringing-forth [*Pro-duire*] comes to pass
> only insofar as something concealed comes into unconceal-
> ment. . . . Technology is therefore no mere means. Technol-
> ogy is a way of revealing. If we give heed to this, then another
> whole realm for the essence of technology will open itself up
> to us. It is the realm of revealing, i.e., of truth [*Wahr-heit*].[22]

Let us jump ahead so as to note straight away the difference
between this technology and modern technology: the essence of
modern technology is Enframing (*Gestell*) as a mode of unconcealment, a particular mode which paradoxically obscures unconcealment. That which leads us to question the essence of technology
is the thing that both reveals and conceals that essence: "the
unconcealment in accordance with which nature presents itself
as a calculable complex of the effects of forces can indeed permit
correct determinations; but precisely through these successes
the danger can remain that in the midst of all that is correct

22 Ibid., 11–12. One page later, Heidegger writes: "It is as revealing, and not as
manufacturing, that *technē* is a bringing-forth."

the true will withdraw."[23] This is still only a temporary formu-
lation of the paradox mentioned above, but already at this stage
we should ask how it is justified. First of all by the fact that it is
specific to modern technology that is does not unfold "into a
bringing-forth [*une pro-duction*] in the sense of *poiēsis*. The revealing that rules in modern technology is a challenging [*Heraus-
fordern*], which puts to nature the unreasonable demand that it
supply energy that can be extracted and stored as such."[24]And
secondly by the fact that the mode of unconcealment that is
Gestell as the essence of modern technology makes technology
appear as a scientific application, so concealing the fact that it
makes scientific "exactitude" possible:

> Because the essence of modern technology lies in Enframing,
> modern technology must employ exact physical science.
> Through its doing so, the deceptive illusion arises that
> modern technology is applied physical science. This illusion
> can maintain itself only so long as neither the essential
> origin of modern science nor indeed the essence of modern
> technology is adequately found out through questioning.[25]

To this is added the further illusion whereby man only ever
encounters himself and his works, while Enframing is not his
doing but what calls him forth, reducing man himself to a stand-
ing reserve. So, in Enframing (*Gestell*), "unconcealment" as destiny
(*Geschick*) comes about as a danger or peril (*Gefahr*). But this is
ultimately due to the fact that in Enframing, as the "essence" of
modern technology, the retreat of unconcealment carries the
day, dissociating the truth of Being from itself in its essence—a
formula whose meaning we will need to clarify:

> Where Enframing holds sway, regulating and securing of the
> standing-reserve mark all revealing. They no longer even let

23   Ibid., 26.
24   Ibid., 14.
25   Ibid., 23.

their own fundamental characteristic appear, namely, this revealing as such.

Thus the challenging Enframing not only conceals a former way of revealing, bringing-forth, but it conceals revealing itself and with it That wherein unconcealment, i.e., truth, comes to pass.

Enframing blocks the shining-forth and holding-sway of truth. The destining that sends into ordering is consequently the extreme danger. What is dangerous is not technology. There is no demonry of technology, but rather there is the mystery of its essence. The essence of technology, as destining of revealing, is the danger.[26]

But even though it is from itself as essence that the truth of Being is dissociated by Enframing, it is Enframing that bears the "advent" (*Ereignis*) of "another beginning" (*anderer Anfang*), one where, by way of the "essence" of technology, the truth of Being is no longer an essence: "It is technology itself that makes the demand on us to think in another way what is usually understood as 'essence' ['*Wesen*']."[27] This is how Heidegger understands his recourse to Hölderlin's phrase: "But where danger is, grows the saving power also."[28] And it is precisely at this point that the Heideggerian thinking of technology opens onto another non-anthropology, which is no doubt truer to itself because it has dispatched any "destiny" bound to the "essence of human being": the non-anthropology of Simondon's thinking of human being and technology.

Let us be clear. If the reign of technology is the last epoch in the history of Being itself inasmuch as it does not reveal itself except in its retreat, it is still the case that this terminal unconcealment

26   Ibid., 27–28. See also "The Turning": Heidegger declares that when the danger has been brought to light, then, as we shall see, the exit from metaphysics also becomes possible.
27   Heidegger, "The Question Concerning Technology," 30.
28   Ibid., 28.

of Being marks the accomplishment of metaphysics as the objectification and forgetting of Being:

> The world changes into object. In this revolutionary objec-
> tifying of everything that is, the earth, that which first of all
> must be put at the disposal of representing and setting forth,
> moves into the midst of human positing and analyzing. The
> earth itself can show itself only as the object of assault, an
> assault that, in human willing, establishes itself as uncondi-
> tional objectification. Nature appears everywhere—because
> willed from out of the essence of Being—as the object of
> technology.[29]

The anti-metaphysical character of the thinking of *Gestell* involves understanding this accomplishment of metaphysics in *Gestell*. There is in this accomplishment a fundamental ambiguity—on the one hand, *Gestell* completes the objectification of being and the forgetting of Being as it conducts the object (*Gegenstand*) towards a "standing-reserve" (*Bestand*); but on the other, because the *ob*-ject defines *Vorhandenheit*, the "standing-reserve" which extends it already and necessarily gestures outside *Vorhand-enheit*, and is *"zuhanden"* and revelatory of being-in-the-world, as well as being the accomplishment of metaphysics. The fundamen-tal ambiguity of the "standing-reserve" is evident in this passage: "Yet an airliner that stands on the runway is surely an object. Certainly. We can represent the machine so. But then it conceals itself as to what and how it is. Revealed, it stands on the taxi strip only as a standing-reserve, inasmuch as it is ordered to ensure the possibility of transportation."[30] The means of transport only differs from the object because it is also a reference and not just

29 Heidegger, "The Word of Nietzsche: 'God is Dead,'" in *The Question Con-cerning Technology and Other Essays*, 100.

30 Heidegger, "The Question Concerning Technology," 17. To observe the ambiguity of Heidegger"s thinking of the subject-object relation, one need only read "Science and Meditation." Here it is the ambiguity of "reserve" (*Bestand*), simultaneously the accomplishment of metaphysics and the exit from the "object" which characterized it.

a means in which the essence of technology is radically forgotten. In other words, *Gestell* as the essence of modern technology itself reveals here what only *Sein und Zeit* had previously revealed: the irreducibility of being to the *Vorhandenheit* of the object, which is to say the system of reference that is the world as a complex of "tools," and which the "artisanal instrument"[31] is incapable of revealing *because* that is what it is. The fact that Heidegger says here that the artisanal instrument is "independent," opposing it in this way to the "absolutely dependent"[32] modern machine, does not invalidate our interpretation, but rather confirms that modern *Gestell* produced *Sein und Zeit* itself: not that Heidegger had denied his own thought in the meantime,[33] but that the forgetting of forgetting that is *Gestell* bears within itself the thinking of Being, inaugurated in *Sein und Zeit*, as "that which saves," because the modern machine can no longer conceal itself as "reference."

So, if Enframing, the "essence" of modern technology, is a destining of unconcealment which has become a "danger" in that it has withdrawn completely, then this is ultimately because the aspect of the withdrawal constitutive of any unconcealment is absent, revealing that the truth of Being is not essence, a revelation that is the pure withdrawal constitutive of this absence of withdrawal.[34] Such is the non-self-identity of the truth of Being

31   Ibid., 17.
32   Ibid.
33   On this point see the "Letter to Richardson," in William Richardson, *Heidegger: Through Phenomenology to Thought*, 4th ed. (New York: Fordham University Press, 2003).
34   On the "identity crisis" of Being at the end of metaphysics, see Michel Haar's very fine text, "Le tournant de la détresse," in *Cahier de l'Herne: Heidegger*, 335–36. On what we will call the "antinomies" of the thinking of Being, see Heidegger's *The Fundamental Concepts of Metaphysics*, trans. William McNeill and Nicholas Walker (Bloomington: Indiana University Press, 1995). This text, which is without doubt one of Heidegger's most fundamental, prefigures—conceptually and so aporetically rather than poetically—the final "tautological thinking." The return to Anaximander made by Heidegger after this course in texts such as "The Anaximander Fragment" (in *Early Greek Thinking*, trans. David F. Krell & Frank A. Capuzzi [New York: Harper &

qua non-question, and the Heideggerian thinking of the essence   **63**
of technology—as *Gestell* bearing *Ereignis*—is what allows philosophy to take leave of the (non-)question of Being. By which we
understand that in the end this (non-)question could only come to
what we will call its self-transcending sense by finally uncovering,
in the fundamental ambiguity of *Gestell* bearing *Ereignis*,[35] its own
metaphysical—anthropological, even—unthought: the deter-
mination of "technology" as something instrumental and human,
which (ontologically) differs from the "essence of technology."
If, on the contrary, Heidegger had made an initial distinction
between use—of means by human being—and fabrication-oper-
ation—he would not have reduced technology in this way and

Row, 1975], 13–58) could be interpreted as a falling back of the Heideggerian
history of Being into a posture that is only non-Hegelian in a Hegelian way,
and which consists of absorbing the history of Being into the "beginning"
that was unrecognized by Hegel. Not that I, for my part, do not recognize the
difference that Heidegger, starting from this Hegelian non-recognition of the
"beginning," means to indicate between himself and Hegel "with respect to
the intention of thought, with respect to the law and character of a dialogue
with the history of thought"—which is the difference between the *Aufhebung*
and the "step back" (see *Identity and Difference*, trans. Joan Stambaugh
[Chicago: University of Chicago Press, 1969], 49). "Thinking recedes before
its matter, Being, and thus brings what is thought into a confrontation in
which we behold the whole of this history—behold it with respect to what
constitutes the source of this entire thinking, because it alone establishes
and prepares for this thinking the area of its abode. In contrast to Hegel, this
is not a traditional problem, already posed, but what has always remained
unasked throughout this history of thinking." (Ibid., 50) However, apart
from the fact that we have here a reading of Hegel which Heidegger himself
would, in other texts, authorize us to challenge, it is not certain that it is possible to escape Hegel as long as you lay claim to the "source" of the history of
thinking. In order to examine this point, I could refer to a polemical revival of
the French interpretation of Heidegger, precisely as concerns his problematic relationship with Hegel. In writing this I am thinking in particular of the
work of Christian Ferrié and François Raffoul.
35   More precisely, "What we experience in the Enframing as the constellation of
Being and man through the modern world of technology is a *prelude* to what
is called the event of appropriation [*Er-eignis*]" (*Identity and Difference*, 36,
author's emphasis). [Translation slightly modified. —Trans.]

would not have relied on the ontological difference to save the *essence* of technology from this supposedly ontic sphere.

From Possible Dialogue to Inevitable Misunderstanding: The Self-Transcendence of Heidegger's Questioning and Simondon's Unthought

This internal critique of Heidegger's thinking of technology should not lead us to think that Simondon has brought the matter to a conclusion. If it is true that one may discern an auto-transcendent meaning in Heidegger's thought, as I have been able to elsewhere with respect to Husserl,[36] then the metaphysical and anthropological unthought in Heidegger's thinking is only the other side of a questioning to come, borne already by this thought and whose depths exceed Simondon's ontogenetic problematic. This is what I must now elucidate. It is my conviction that the two problematics, Simondon's and the one whose simple possibility is indicated in Heidegger's thinking, can be articulated on the basis of the internal critique of Heidegger's thinking as it presents itself—and not as it would like to present itself: that is, as radically non-objectivizing.[37] Just like the question of non-anthropology, the question of non-objectification is in fact a question that brings Heidegger and Simondon together. But it makes the first into the precursor of a radically non-objectivizing problematic for which Simondonian ontology furnishes a secondary translation—secondary because less profound, even if appropriate. In order to understand this, I am going to start from an unresolved paradox from the preceding discussion. Here is the paradox: it has emerged that Heidegger's

36 See my article, "Husserl et l'auto-transcendance du sens," *Revue philosophique de la France et de l'étranger*, no. 2 (2004). On the concept of the auto-transcendence of sense, see also my book *Penser l'individuation*, Introduction, 2.

37 Here I use "objectivizing" rather than "objectifying" in order to express *my own* thought. On the English translation of Heidegger on this score, see n. 5.

thinking of technology is in a sense more anthropological than Simondon's, and yet it is also in a sense more destinal—destiny not being understood here as that imposed on human being by technology, but as that imposed on human being by his own essence. Now, this destinal thinking is only really what it is because, we remarked above, the essence of human being is no longer understood as an essence in which human being would belong to himself—from which arises the concept of "Being-there" (*Dasein*). But to say that the essence is no longer strictly speaking essence, is to prepare the exit from this anthropology which until this point had been paradoxically reconciled with destinal thinking. Conversely, Simondon's non-anthropology has also ultimately emerged as relatively destinal: technics comes to shape a civilization through a process of "concretization" which makes it auto-conditioning. There is no paradox here but, on the contrary, a very logical association between non-anthropology (assumed this time) and destinal thinking. Now, Heideggerian destiny differs from Simondonian destiny because for Heidegger it is ultimately neither technology nor human being that is destined, but Being. And yet the fact that the question of Being proves to be a non-question does not in any sense indicate that there could not be a question more radical than Simondon's ontogenetic question—a more radical question which, if based on an internal critique of Heideggerian thought, may well lead this time, in a second instance, to Simondon's ontogenetic and non-anthropological thinking, as though to both validate it and put it into perspective, all the while liberating it from the destinal character that burdens it. It is this point that I would like to clarify in conclusion.

Even if the internal critique of the Heideggerian thinking of *Gestell* seems to vindicate Simondon while rebuking Heidegger for staying within an anthropological thinking of technology, there nevertheless remains what we highlighted right at the beginning: in Heidegger's view, Simondon would, for his part, adhere to an anthropological reduction of the essence of human being to a being-present-at-hand. Not that this rebuke can stand as it is, since it is made in the name of an essence of human being which is now problematic. But the accusation of a reduction to being-present-at-hand is doubtless valid beyond the debate over the essence of human being: the reduction to being-present-at-hand is not characterized as a particular thesis that Heidegger would decry, but as a general attitude of the philosophizing individual himself.[38] Now, Simondon's ontogenetic problematic may well remain in keeping with this attitude, to the extent that what Heideggerian "ontological difference" names is the exit from this kind of attitude by way of a double phenomenological reduction leading—beyond Husserl's still egological intentionality—to being-in-the-world; while Simondon, for his part, makes no reduction—except, maybe, a Bergsonian "reduction to becoming."[39] To emphasize: speaking of a "double reduction" with respect to Heidegger does not mean that he remains within the phenomenological sphere as defined by Husserl, but that the thinking of being-in-the-world may be understood as a "reduction" which comes to limit the pretentions of the first reduction while benefitting from the "step back" inherent in what it thus limits: "fundamental" ontology is the heir to phenomenology in its distinction from ontology. It is this distancing that Simondon's properly ontological, even cosmological, approach, which derives more from Bergson, does not possess.

Of course, genetic ontology or "ontogenesis," is characterized, as Simondon says, by a certain "ontological difference," which in addition indicates once more a way out of the objectification of being. At issue is the difference between the individual and pre-individual reality—the latter, incidentally, referred to by

---

38 On the limits of this archi-reflexive questioning in Heidegger, see my article "Hegel et l'impensé de Heidegger," *Kairos*, no. 27 (2006): 89–110.

39 The expression is taken from Merleau-Ponty's "Bergson se faisant," *Bulletin de la Société française de philosophie*, no. 1 (1960): 35–45.

Simonond as "Being *qua* Being".[40] In any case, Simonond's conceptual configuration is neither clear nor radical and self-sufficient.
Firstly, it is not clear because individuation, which is the non-*ob*-
ject, is both the same as and different to the pre-individual as
it differs from the individual. And it is not radical and self-sufficient, but based on what Simonond calls "schemas of physical
thinking."[41] In fact, and more profoundly still, the way in which
Simonond anchors his approach in Bergsonian and Bachelardian
themes indicates that he misses the primacy of the anti-foun-
dational and radical question of sense. It is certainly possible to
say that Simonond develops the genetic and anti-substantialist
ontology that is the counterpart to Bachelardian epistemology.[42]
But it is precisely because of this that he prevents himself from
giving his questioning of the subject-object relationship the necessary depth and reflexivity that would allow it to catch sight of
the paradoxical constitution of the subject by the object understood as sense—a paradoxical constitution that in fact only
a double reduction reversing "natural" or naive intentionality
allows us to glimpse.

Even if the Heideggerian question of Being cannot be completely
identified with this question of sense either, it leads to it, at least

40 Simonond, *L'individuation à la lumière des notions de forme et d'information*,
    317. In the Introduction of the same book, Simonond uses the expression
    "Being as it *is*," which indicates that what he calls "Being" (*Sein / être*) would
    be a being (*Seiende / étant*) from Heidegger's point of view: in Heidegger,
    Being *isn't*, but "there is" Being (*es gibt Sein / il y a l'être*).
41 Ibid., 327–28. The "biological" or "technical" (*ibid.*) schemas of thinking are in
    fact based on these physical schemas, in Simonond, thanks to *contemporary*
    physics, which broadens physical rationality. That is why the conclusion of
    the book only develops the question of the physical schemas.
42 On this point see my book *Simonond ou l'encyclopédisme génétique* (Paris:
    PUF, 2008), 9–13, as well as my article "D'une rencontre fertile de Bergson et
    Bachelard: l'ontologie génétique de Simonond," in *Bergson et Bachelard: Con-
    tinuité et discontinuité*, ed. Frédéric Worms and Jean-Jacques Wunenburger
    (Paris: PUF, 2008), 223–38. On Bachelard's constancy, pertinence, as well as
    his limits regarding his relation to Husserlian phenomenology, see Bernard
    Barsotti's *Bachelard critique de Husserl* (Paris: L'Harmattan, 2003).

potentially, precisely at the high point of the interrogation of non-objectivity through the thematics of "worldhood" and "meaningfulness" (*Bedeutsamkeit*). In fact, it is in §§ 12–18 of *Sein und Zeit* that, in the first place, being-in-the-world proves to be irreducible to an object of knowledge, knowledge being rather only a mode of being-in-the-world. At the same time it is revealed that *praxis* is still inherent in *theoria*, the latter being a mode or dimension of a *praxis* which, when "ontologized" in a non-materialist way— such are the commonalities and the difference between Marx and Heidegger—is equivalent to being-in-the-world because it is multi-modal or multi-dimensional. Indeed, there is a kind of multi-dimensionality to being-in-the-world, a multi-dimensionality whose establishment the thematic of "meaningfulness" would have permitted had Heidegger not made it into a simple system of "reference" instead of multi-dimensionally diffracting every signification—here understood as representation ("tree," "table," "concept," etc.). But, multi-dimensionally diffracting every signification is the same as no longer speaking of anything but the sense *that makes me*, and it is certainly here that Heidegger—who was less interested in giving a new meaning to Thales's "know thyself" than in revisiting the thinking of Being—was unable to accept the consequences of his questioning, unless in the later form of a tautological thinking seeking to *say* something without speaking *about* something. So this new questioning that I have in mind radically interrogates the attitude of the philosophizing individual himself. It is the multi-dimensional diffraction of significations that prevents the philosophizing individual from continuing to reduce significations to identities of *ob*-jects of the mind, and therefore from continuing to unknowingly absolutize himself as someone not-constituted by the sense "present-at-hand," and so as originary or a "*subjectum*": the multi-dimensional diffraction of significations would allow the philosophizing individual to adopt a completely anti-natural *attitude* which would not contradict the *thesis* of this philosophizing individual on the finitude of *Dasein* as constituted by being-in-the-world.

Now, while it is not possible here to effect the multi-dimensional
diffraction of significations so as to elaborate the new radically
non-objectivizing problematic, it is at least possible to indicate
how this internal overturning of Heideggerian questioning brings
about an all-encompassing relativization—which is also to say a
validation—of Simondon's genetic ontology, where the only error
was to think of it as "first philosophy."[43] The difference between
multi-dimensional sense and the object of knowledge *which is
certainly a dimension of sense—but only one dimension*—translates (internally, this time, to this sole dimension of sense that
is the object of knowledge) into the difference between individual and substance. Now, it is precisely the difference between
individual and substance that is foundational for Simondon's
genetic ontology: knowledge of individuation is knowledge of
beings as relations and not substances. If this knowledge is
also individuation of knowledge, and genetic ontology claims to
be first philosophy and not a secondary translation, however
appropriate, of another problematic, then Simondon certainly
needs the knowledge of individuation to be a non-objectivizing
knowledge. But it is contradictory to claim a knowledge that is
at the same time non-objectivizing, and Simondon in fact continues to objectivize significations, manipulating them to make
them equal to what he is speaking about, instead of thinking of
sense as individuating itself in him. And another contradiction
goes hand in hand with this one: Simondon attributes to the
knowledge of individuation the privilege of being individuation
of knowledge, but he at the same time affirms that all knowledge

43   For an account of the difficulties Simondon encounters here, as well as
      the resolution of these difficulties brought by the "all-encompassing rel-
      ativization" of his ontology within a System of Philosophical Relativity now
      opened by a "philosophical semantic" (one in which the philosophizing
      individual only claims to think the sense *by which he is produced*), see the
      last chapter of my book *Penser la connaissance et la technique après Simondon*
      (Paris: L'Harmattan, 2005). The program of Philosophical Relativity is also
      expounded, more pedagogically, in my article "Penser après Simondon et
      par-delà Deleuze," in *Cahiers Simondon No. 2*, ed. Jean-Hugues Barthélémy
      (Paris: L'Harmattan, 2010).

is individuation of knowledge and that the theory of knowledge must not precede ontogenesis.[44] These contradictions reveal that knowledge of individuation remains a knowledge and can only claim to be non-objectivizing because it *appropriately translates*, in the sole ontological dimension, the radically non-objectivizing attitude which the philosophizing individual must adopt in the *first* problematic—a first problematic which should multi-dimensionally diffract every signification, only then to translate itself in each of the dimensions of sense thus released, and rediscover, in one of these uni-dimensional translations, Simondon's genetic ontology and the truth of his non-anthropology.

Works Cited

Barsotti, Bernard. *Bachelard critique de Husserl*. Paris: L'Harmattan, 2003.

Barthélémy, Jean-Hugues. "D'une rencontre fertile de Bergson et Bachelard: l'ontologie génétique de Simondon." In *Bergson et Bachelard: Continuité et discontinuité*, edited by Frédéric Worms and Jean-Jacques Wunenburger, 223–38. Paris: PUF, 2008.

———. "Hegel et l'impensé de Heidegger." *Kairos*, no. 27 (2006): 89–110.

———. "L'humanisme ne prend sens que comme combat contre un type d'aliénation" (interview with Ludovic Duhem), *Tête-à-tête*, no. 5 (2013): 54–67.

———. "Husserl et l'auto-transcendance du sens." *Revue philosophique de la France et de l'étranger*, no. 2 (2004): 181–97.

———. "Penser après Simondon et par-delà Deleuze." In *Cahiers Simondon No. 2*, edited by J.-H. Barthélémy, 129–46. Paris: L'Harmattan, 2010.

———. *Penser la connaissance et la technique après Simondon*. Paris: L'Harmattan, 2005.

———. *Penser l'individuation: Simondon et la philosophie de la nature*. Paris: L'Harmattan, 2005.

———. *Simondon*. Paris: Belles Lettres, 2014.

———. *Simondon ou l'encyclopédisme génétique*. Paris: PUF, 2008.

———. "What New Humanism Today?" Translated by Chris Turner. *Cultural Politics* 6, no. 2: 237–52.

Haar, Michel. "Le tournant de la détresse." In *Cahier de l'Herne: Heidegger*, edited by Michel Haar, 335–36. Paris: Editions de l'Herne, 1983.

Heidegger, Martin. "The Anaximander Fragment." In *Early Greek Thinking*. Translated by David F. Krell and Frank A. Capuzzi, 13–58. New York: Harper & Row, 1975.

44 See Simondon, *L'individuation à la lumière des notions de forme et d'information*, 36, 264 and 284.

———. *The Fundamental Concepts of Metaphysics*. Translated by William McNeill and Nicholas Walker. Bloomington: Indiana University Press, 1995.

———. *Identity and Difference*. Translated by Joan Stambaugh. Chicago: University of Chicago Press, 1969.

———. "Letter to Richardson." In William Richardson. *Heidegger: Through Phenomenology to Thought*. 4th ed., VIII–XXIII. New York: Fordham University Press, 2003.

———. "The Question Concerning Technology" In *The Question Concerning Technology and Other Essays*. Translated by William Lovitt, 3–35. New York: Harper Perennial, 1977.

———. "The Turning." In *The Question Concerning Technology and Other Essays*, 36–49.

———. "The Word of Nietzsche: 'God is Dead.'" In *The Question Concerning Technology and Other Essays*, 53–112.

Merleau-Ponty, Maurice. "Bergson se faisant." *Bulletin de la Société française de philosophie*, no. 1 (1960): 35–45.

Simondon, Gilbert. *Du mode d'existence des objets techniques*. Paris: Aubier, 1958.

———. *L'individuation à la lumière des notions de forme et d'information*. Grenoble: Millon, 2005.

Taminiaux, Jacques. "L'essence vraie de la technique." In *Cahier de l'Herne: Heidegger*, edited by Michel Haar, 263–84. Paris: Editions de l'Herne, 1983.

Publication Details and License Information

Aspects of a Philosophy of the Living

Original title: "Individuation et apoptose: repenser l'adaptation? Simondon, le vivant et l'information."

Published in: *Nouvelles représentations de la vie en biologie et philosophie du vivant*, edited by Jean-Claude Ameisen and Laurent Cherlonneix.

Technology and the Question of Non-Anthropology

Original title: "La question de la non-anthropologie."

Published in: *Technique, monde, individuation: Heidegger, Simondon, Deleuze*, edited by Jean-Marie Vaysse.

Lightning Source UK Ltd.
Milton Keynes UK
UKOW02f1438010616

275379UK00002B/34/P